Forests come to life in spring

JILL KALZ

Hardcover edition published by Creative Education

123 South Broad Street, Mankato, Minnesota 56001

Creative Education is an imprint of The Creative Company

RiverStream Publishing reprinted by arrangement with The Creative Company

Designed by Rita Marshall

Photographs by Corbis (W. Perry Conway), Dennis Frates, Getty Images (Larry Dale Gordon, Darrell Gulin), Tom Stack & Associates (Sharon Gerig, Thomas Kitchin, Merrilee Thomas), The Viesti Collection

Cover illustration © 1996 Roberto Innocenti

Copyright © 2006 Creative Education

International copyright reserved in all countries. No part of this book may be reproduced in any form without written permission from the pubisher.

Printed in the United States of America

Library of Congress Cataloging-in-Publication Data

Kalz, Jill. Spring / by Jill Kalz

p. cm. — (My first look at seasons)

ISBN 1-58341-363-4

1. Spring—Juvenile Literature. I. Title.

QB637.5.K35 2004   508.2—dc22   2004056243

1 2 3 4 5 CG 15 14 13 12

RiverStream Publishing—Corporate Graphics, Mankato, MN—112012—1005CGF12

Spring

Good-bye to Winter 6

Time to Wake Up 12

Eggs and Babies 14

A Colorful Season 18

Hands-On: Taco in a Bag 22

Additional Information 24

Good-bye to Winter

March is here! Snow melts. Many birds fly north from their winter homes. Flowers bloom. In the northern half of the world, March 20 is the first day of spring.

Spring is one of Earth's four **seasons**. The other seasons are summer, fall, and winter. Each season lasts about three months. Spring comes between winter and summer.

Flowers grow and open up in spring

It usually rains a lot in spring. Light rains are called showers. Rain helps flowers grow. People sometimes say, "April showers bring May flowers."

Spring is a stormy time. When it rains too much, rivers and lakes **flood**. Spring may bring tornadoes, too. Tornadoes are windy storms that spin across the land.

Some people say a tornado sounds like a train.

Others say it sounds like a roar.

Tornadoes are strong and scary storms

Many flowers grow next to this stream

Time to Wake Up

Plants need sunlight and water to grow. Spring gives plants a lot of both. When the ground warms up and spring rain falls, seeds **sprout**.

Many plants go to sleep in winter. When spring comes, they wake up. Some plants wake up earlier in spring than others do. Tulips bloom early. So do apple trees and plum trees.

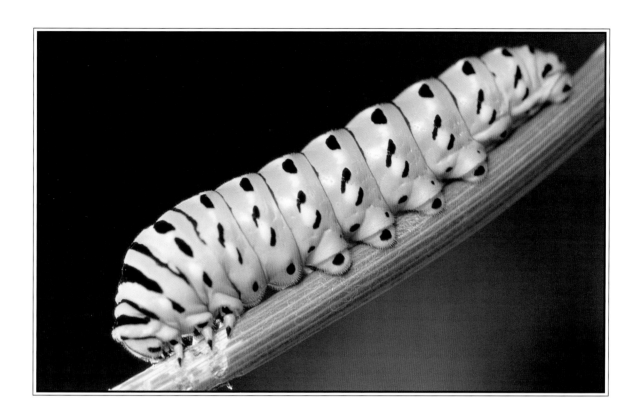

On the first day of spring,

day and night are each

12 hours long.

CATERPILLARS HATCH IN SPRING AND THEN FIND FOOD

Trees and bushes grow small bumps on their branches in spring. These bumps are called buds. Buds turn into leaves or flowers. Roses, daisies, and other plants grow buds, too.

Eggs and Babies

Hibernating animals wake up in spring. Turtles crawl out of the mud to warm themselves in the sun. Chipmunks, gophers, and snakes wake up, too.

Robins are one of the first birds people see in spring. They are called "messengers of spring."

A MOTHER ROBIN FEEDING HER YOUNG CHICKS

Many baby animals are born in spring. Spring is a good time to be born. Food is easy to find, and the sun feels warm.

Robins and other birds fly north in spring. They sing songs to each other. They build nests. They lay eggs and wait for their babies to hatch.

SPRING IS A SEASON OF BABIES AND NEW LIFE

Insects lay eggs, too. Flies and beetles lay eggs. Spiders lay eggs in small bags called egg sacs. Moths and butterflies break out of shells called cocoons in spring.

A Colorful Season

Spring is planting time. Farmers plant seeds in fields. Some people plant seeds in vegetable gardens. Others plant flowers in pots and boxes.

This big flower is called an Easter lily

Spring is filled with **holidays**. Families gather to pray and eat special foods during Passover and Easter. Some people color eggs on Easter. April Fool's Day is a spring holiday. On April 1, people play tricks on each other. Mother's Day and Memorial Day are spring holidays, too.

Spring is a colorful season. Enjoy it! Plant seeds! Whistle with the baby birds! Color Easter eggs! And then, get ready for summer!

The average American
eats 250 eggs each year.
This includes eggs in
cakes and other foods.

EASTER EGGS CAN BE FANCY AND COLORFUL

Hands-on: Taco in a Bag

Celebrate Cinco de Mayo by cooking for your family!

What You Need

One pound (.45 kg) ground beef

One packet taco seasoning mix

Four snack-size bags of corn chips

Toppings such as shredded lettuce, shredded cheddar cheese, and salsa

What You Do

1. Have a grown-up help you brown the meat. Then, follow the directions on the seasoning packet.
2. Before opening the bags, gently crush the chips.
3. Have a grown-up help you cut the bags along a side edge. Spoon the beef and other toppings into the bags. Grab a fork and eat!

Many people have parties on Cinco de Mayo

Index

animals 14, 16
birds 6, 15, 16, 20
farmers 18
holidays 20, 22
insects 18
plant buds 14
plants 12, 14, 18
rain 8
tornadoes 8, 9

Words to Know

flood—fill with too much water

hibernating—being in a very deep sleep for weeks or months

holidays—special days that happen every year

seasons—the four parts of a year: spring, summer, fall, and winter

sprout—grow; when a seed sprouts, it grows roots and a stem

Read More

Fisher, Aileen. *The Story of Easter*. New York: HarperCollins Children's Books, 1998.

Sper, Emily. *Passover Seder*. New York: Scholastic, 2003.

Thayer, Tanya. *Spring*. Minneapolis: Lerner Publishing Group, 2001.

Explore the Web

FEMA for Kids: Tornadoes http://www.ready.gov/know-facts

Note to Parents, Teachers and Librarians: We routinely verify our Web links to make sure they are safe and active sites. So encourage your readers to check them out!